Number Poetry

by Ben Ditmars

Copyright © 2017
Benjamin Ditmars

All Rights Reserved

No part of this book may be reproduced, stored in a retrieval system or transmitted in any form or by any means without the prior written consent of the author, except by a reviewer who may quote brief passages in a newspaper, magazine or journal.

Table of Contents

Greater Detail

Jack Used the Number Line

Fiendfyre

Solar Radii

Shameless Edges

The Magnitude of Hate

Echoes of Laughter

Tangent Line

Bright as Death

Shadows

D-20

Mutant Love

Bought Cheap

Synonyms for Love

Drums

Burned Impulse

Paint Your Dream

Fantasists

Echoes into Hell

Greater Detail

explain to me in
greater detail why I count
down the stars - they're gone.

Jack Used the Number Line

Jack used the number line to solve.

Write a letter to Jack telling him what did he did wrong and how to fix his mistake…

Jack, difference is the result of a subtraction problem. The remainder left after subtraction of *1* value from another.

We have something we must lose; years or sanity

Quite often isolated numbers counted in the miles we've spent searching.

But I will tell you there are no mistakes and fixing what is right

Too often tricks us into doing wrong.

Trust me. You'll find out down the line.

Fiendfyre

I saw the will to change
Extinguished like a *1,000*
Beasts rose high, like chills
When I could see her and
Remain invisible.

Solar Radii

I've spent time measuring
a volume of kind words,
aggression spilled on
pavement equivalent the
mass of ugliness

he didn't stand a chance as
bullets clashed with liters of
intolerance and hit between the
solar radius of *3* young hearts.

Shameless Edges

Consensual and
shameless edges proven
wrong while we were
forest bathing
and the same.

The Magnitude of Hate

the most massive star will
live a million years
before white knuckles and
a flash of light

reckoning we might see
coming as we will gain
4,000 x the magnitude
of hate.

Echoes of Laughter

"It's all accumulation and the aftermath," she says as I would question her about the failing earth and giants unaware that they are sinking everyday

We share what's left of predetermination as the echoes of her laughter warm my heart.

Tangent Line

I did the math while switching stations to determine at what point

We became a tangent line touching at the curve of an ellipse and never crossing.

Bright as Death

her explosive life-force
bright as death could
live *1,000* years

surviving warlike multiplicity

a waxing moon and
thoughts

near termination.

Shadows

Shadows are the shade within
clear boundaries; deep desired
whispers we have found

Imagine artificial light and you
will see our lust corroded with
despair,

The cotton fabric of existence
torn as waves crash down.

D-20

Glowing numbers
Rolling red like
Dragon fire in a
Cave where we
Keep warm.

Mutant Love

They say your innocence is lost

damage, errors in your
replication degenerate a sterile
nature

but you no less strong, ungodly
or unrighteous.

Inheritance from sin becomes
experience discernable in

cytoplasmic fragments of a
stolen past.

Bought Cheap

"I deserve hell," I exclaimed
into the bathroom mirror
realizing my jeans were sewn
in hell and I had bought them
cheap

There are more quotes on the
internet about food than
poverty, but that can't change
reality

there are songs about life
but fewer about death and that
is some small consolation

as there are many more on love
than there are hate.

Synonyms for Love

1,025,109 English words
47 synonyms for love,
18 you have heard before
13 you will hear again

8 PM arrives too soon
6 AM remains elusive

4 nights and the
2 of us see
1

Drums

Drums sometimes break through
Rhythm and they play alone
Heartbeats tapping glass.

Burned Impulse

My impulse missed a beat
or *2*, submerged in life
the scent of perfume and
fluorescent lights burned
beige from overused
transcendence.

Paint Your Dream

Inspired by the words of Jacqueline E. Smith

If you use the same paint, then
it all works out and sometimes
it's necessary to let the paint
dry, because life is painting
and takes time –

Your hands are red acrylic
bleeding ink, but I take them in
mine and blend.

Fantasists

We became those people overnight
Fantasists who day daydream
holding hands and speak in
whispers of true love.

Like nothing came before a kiss
across the console of a Chevy
Cruz or will exist without space
curvature across the low points
of our spine.

I've fanned the flames and
Lost myself in your geometry.

Echoes into Hell

we rode fine lines
like they were echoes
into hell and momentary
2 dog nights cast cold
spells where a warmth
resided.

Ben Ditmars was first published in his college publications the Cornfield Review and KAPOW and has since been featured in several online literary journals. Currently, he lives in Marion, Ohio where he works in bookkeeping and accounting.

Photograph by Jacqueline E. Smith

Find out more on his website:

https://benjaminditmars.com

www.ingramcontent.com/pod-product-compliance
Lightning Source LLC
Chambersburg PA
CBHW031522210526
45464CB00007B/3005